START WITH A QUESTION

Can People Feel Sound?

FAQs About Energy

Kylie Burns

A Crabtree Forest Book

Crabtree Publishing
crabtreebooks.com

Notes for Readers and Class Discussion

This book is designed to teach readers about core subject areas, build curiosity, and inspire further investigation. Readers are encouraged to build upon what they already know about the subject and engage in topics they want to learn more about. Here are a few guiding questions to prompt reflection and discussion. Possible answers appear in red.

Before Reading:

Read the title and look at the table of contents. What do you already know about energy?

- *I know that there are different forms of energy.*
- *I know that humans can create energy by burning fuels.*

What would you like to learn about energy?

- *I would like to learn about unusual forms of energy.*
- *I would like to learn how light waves and sound waves work.*

During Reading:

Pause after reading each page or chapter. What questions do you have about what you read? What are you curious to know?

- *I wonder how long it would take for fossil fuels to get used up.*
- *I am curious to know if there are plans to fly helicopters on other planets.*

Make connections to what you are reading. How is this like something you already know?

- *I know that animals have superior senses that help them hunt and survive in the wild.*
- *I know that people have developed more environmentally friendly energy options.*

After Reading:

Recall key details about the book. What was the author trying to teach readers?

- *The author was trying to teach readers that people depend on many different forms of energy.*
- *The author was trying to teach readers that some forms of energy are damaging to the environment.*

How did the images and captions help you understand more?

- *The photographs helped me see how different forms of energy are at work in our world.*
- *The captions helped me understand more details about energy.*

Table of Contents

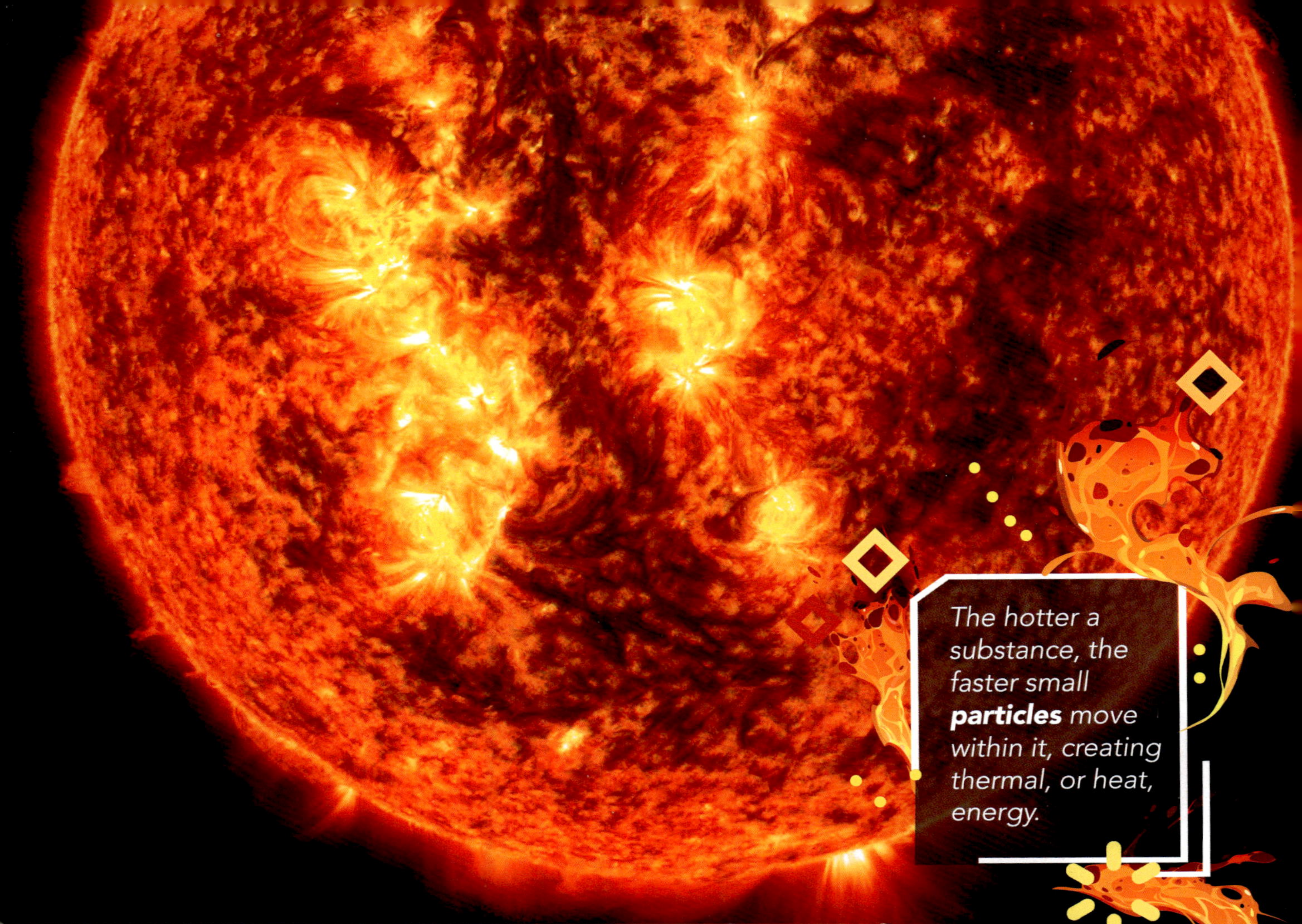

The hotter a substance, the faster small ***particles*** *move within it, creating thermal, or heat, energy.*

Think Like a Scientist

Have you ever wondered where energy comes from? Perhaps you have asked yourself "What makes an electric car work?" or "How do fireflies glow?" In this book, you will discover facts about different kinds of energy that may help answer some of your questions. In fact, asking questions is how scientists begin all of their investigations. They conduct experiments to find answers to their queries, and draw conclusions from the results of the experiments. Sometimes, the results lead to even more questions! Everything we know about energy came from questions and investigations like these.

Everything we do requires energy. In fact, you are probably using several different forms of energy right now. Is the light on? That's electrical energy. Are you holding this book and turning the pages? That's **kinetic** energy. And if you ate something today, you provided your body with energy to move, see, and think!

Read on to find out more about the kinds of energy that fuel our daily lives and power our world. The information you learn may inspire you to ask even more questions about energy—just as a scientist would.

Light energy is created when the Sun's light travels to Earth. It can also come from other sources of energy such as electricity.

Kinetic energy is known as the energy of motion, because if an object is moving, it produces energy.

Is the Sun hotter than lava?

The Sun provides Earth with light and thermal energy, but how does the Sun's heat compare to the temperature of lava?

Extremely hot liquid rock called magma is formed deep under Earth's surface. It rises toward the surface because it is lighter than the rock at Earth's core. When pressure pushes magma up through a volcano, or an opening in Earth's crust, it flows out and becomes lava. As lava flows out of a volcano, it ranges from around 1,300 to 2,200 degrees Fahrenheit (700 to 1,200 degrees Celsius). As it flows, it cools and eventually hardens into rock.

The Sun, on the other hand, is a star made up of burning gas. It is not a liquid, like lava. On the outer surface of the Sun, the temperature is approximately 10,000 degrees Fahrenheit (5,500 degrees Celsius), which is definitely hotter than lava! At the Sun's core, the temperature is a staggering 27 million degrees Fahrenheit (15 million degrees Celsius)!

There are more than 1,500 active volcanoes in the world. An active volcano is one that has erupted in recent history and is likely to erupt again.

DID YOU KNOW?

The Sun may be much hotter than lava, but there is something hotter than the surface of the Sun: lightning! The thermal energy in lightning heats the air to 50,000 degrees Fahrenheit (27,760 degrees Celsius), which is five times hotter than the surface of the Sun. This heat is what causes thunder as the air around the **lightning bolt** vibrates and expands quickly.

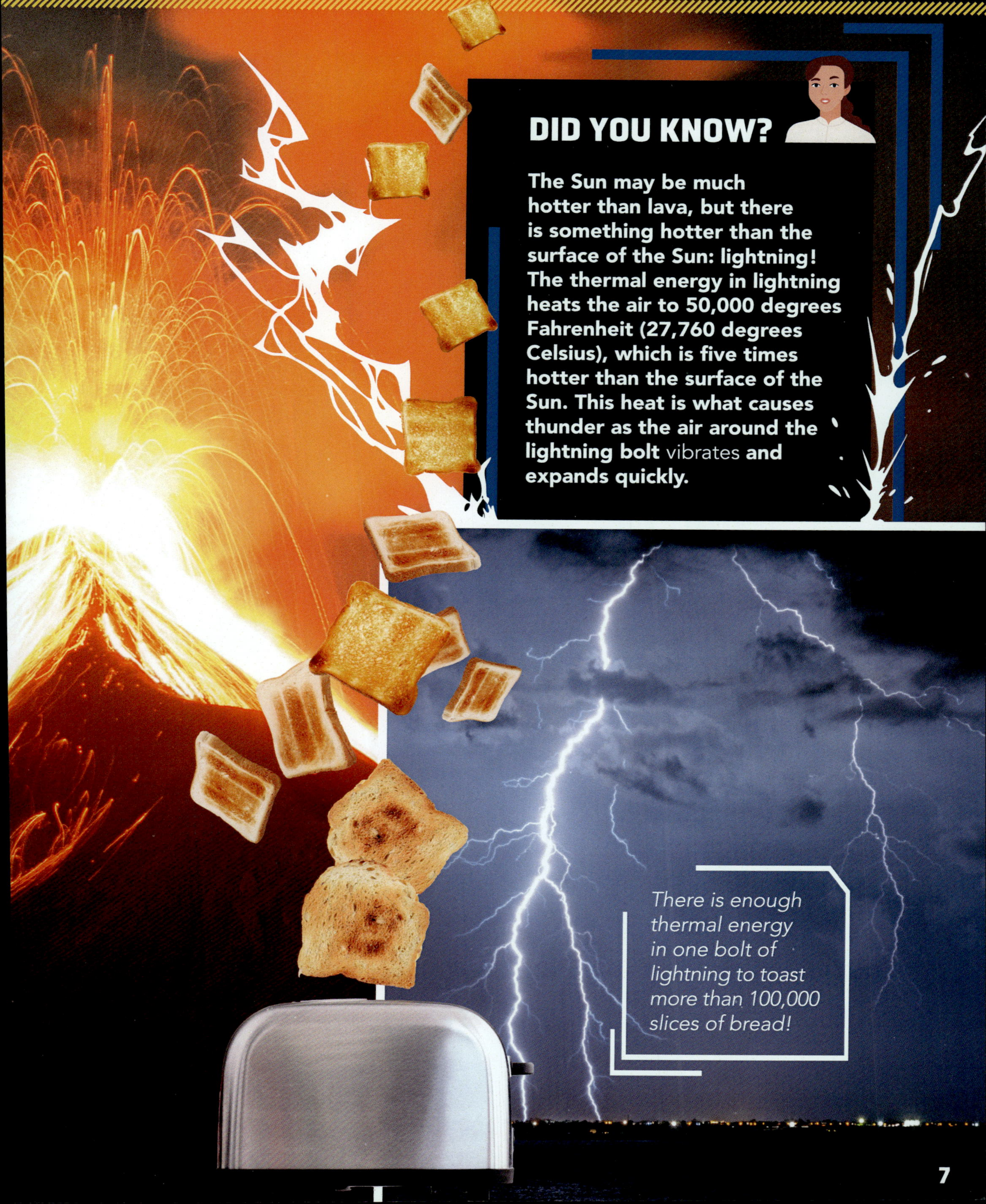

There is enough thermal energy in one bolt of lightning to toast more than 100,000 slices of bread!

Could humans survive without the Sun?

The Sun affects every aspect of life on Earth. Without the Sun's thermal energy, the temperature of Earth's surface would drop below freezing within days. The air and water would freeze, including the surface of the oceans. People could generate some warmth with electricity or by burning **fossil fuels**, but these sources of energy would eventually run out. Without sunlight, it would be completely dark outside. Plants and trees could not survive because they need sunlight for **photosynthesis**. Since plants produce **oxygen** for us to breathe, all other forms of life would soon disappear too, including humans!

*The Sun also affects Earth's **gravity**, keeping it in **orbit** within our solar system. So if the Sun disappeared, Earth would travel through space until it collided with something, such as another planet.*

A hurricane can release as much energy as a 10-megaton nuclear bomb, which could destroy an area half the size of Grand Canyon National Park!

How does the Sun's heat cause hurricanes?

When the Sun heats the surface of tropical ocean water, some water evaporates, or changes from liquid to water vapor in the air. This warm, moist air rises and collides with cooler air, resulting in thunderstorms that form over the surface of the ocean. The thunderstorms create a wall of blowing and swirling wind and rain, resulting in the perfect conditions for hurricanes to form.

Water vapor is water in its gas form.

Can an ice cube have thermal energy?

Thermal energy is found in everything, even in things that are cold! All **matter** is made up of tiny particles called atoms that move faster or slower depending on the temperature. Lower temperatures result in slower-moving atoms, and higher temperatures cause faster-moving ones.

When you add ice cubes to a glass of water at room temperature, the atoms in the water are moving at a faster rate than the atoms in the ice. The water transfers thermal energy to the ice, which then releases atoms from the ice into the water. As this happens, the water gets cooler and the ice heats up and melts. Thermal energy is exchanged between them, resulting in a "cool" drink!

*In a similar way, a glacier has thermal energy. Glacier melting is caused by heat from warmer air above and **meltwater** below.*

How do popcorn kernels pop?

Popcorn kernels contain a seed, water, and a type of **starch** that provides food for the seed. When a popcorn kernel is heated, the thermal energy changes the water to steam, causing the kernels to expand.

Once the temperature reaches 355 degrees Fahrenheit (180 degrees Celsius), the pressure of the steam pushes against the hard shell of the kernel, causing it to burst open and turn inside out.

A kernel's hard outer shell, or hull, is key to the process. It makes sure that water stays inside the kernel as it heats up, building enough pressure for the kernel to burst open. If a kernel's hull is damaged, it will not pop.

What is a heat island?

A heat island is an area where the temperature is higher than in its surrounding areas. Heat islands can have temperatures 1 to 7 degrees Fahrenheit (0.6 to 4 degrees Celsius) higher than surrounding areas during the day, and 2 to 5 degrees Fahrenheit (1 to 2.8 degrees Celsius) at night.

Heat islands can form in **urban** areas, where there are many buildings, roads, and people. These areas have many paved surfaces and fewer plants and trees than **rural** areas. Human-made building materials such as asphalt, concrete, steel, and brick absorb, or take in, more thermal energy than natural surfaces such as trees and grass. This is the reason why a sidewalk feels much warmer than a grassy surface. Plants also absorb water from the ground and release it as water vapor through their leaves, helping to cool the air.

While plants cool the air around them, human-made building materials have the opposite effect. When these surfaces absorb heat, thermal energy can become trapped. This trapped heat can then warm the air to a higher temperature. Human activities in urban areas also cause more heat. Cars on the road, heating and cooling systems, and construction all give off heat, warming the air.

Cities can add more green spaces to help reduce heat island effects. Planting trees helps block the Sun's energy. Covering paved surfaces with plants is also a useful solution. For example, a green roof can help a building, and the air around it, stay cooler.

Cities with narrow streets and tightly packed buildings can block wind flow, causing higher air temperatures.

Why is it dangerous to look at the Sun?

The Sun is 93 million miles (150 million km) away from Earth, yet it has the power to damage our health if we don't protect ourselves. There is so much solar energy coming toward Earth that it would take 44 million power plants to produce the same amount. Most of this energy is light energy that we can see.

If a person looks directly at the Sun, even for a few seconds, the **retinas** in the eyes will be damaged. The tissue in the retina can actually get burned, like your skin might when it gets sunburned. Because the retina does not have pain receptors, a person might not feel the burn until hours or days later.

It's especially important to not look at the Sun through a camera, telescope, or binoculars because these devices can focus the Sun's rays, causing more damage to your eyes.

Sometimes the damage is short term. But if the damage is more severe, a condition called **solar retinopathy** can occur. This often leads to symptoms such as blind spots, blurred vision, and headaches. Most people fully recover, but sometimes vision problems can be long term. So make sure to protect your eyes with sunglasses and never look directly at the Sun!

Even when viewing a solar eclipse, when the Sun is blocked from our view by the orbit of the Moon, it's important to protect your eyes by wearing special eclipse glasses.

Can we capture the power of lightning for electricity?

When lightning strikes, a high amount of energy flows in a very short period of time. In fact, a powerful lightning bolt could provide enough energy to power a small town for an entire day! However, lightning is unpredictable and dangerous, so scientists have not yet found a way to capture its energy. The energy is released with such intensity that it would be difficult to trap it without damaging equipment needed to store it. Perhaps one day someone will find a way to harness the huge amount of energy within lightning to add another **renewable** source of power to our lives.

Lightning can travel 30,000 times faster than a bullet!

Mirages can be seen in the desert too, appearing like a pool of water in the distance.

How does a mirage happen?

Have you ever been in a car and noticed the pavement up ahead looks like it is covered with a puddle of water, even though it hasn't rained? This is called a mirage. It happens when light waves moving from the sky toward the ground are refracted, or bent, by heated air. When the air close to the road's surface is heated, it rises and meets with cooler air. This causes the light to bend, **projecting** the image of the sky onto the pavement and creating a false image of water on the road.

Mirages can also cause objects to appear displaced, or in a different position than in reality.

What makes fireflies glow?

Fireflies are bioluminescent, meaning they are able to make their own light. A firefly's glow is created through a chemical process in special light-producing organs. A firefly controls when its light flashes by sending oxygen into the chemicals in its light organs, resulting in a cool, bright glow. The light is used to attract prey and potential mates. It also sends a warning to predators that the firefly is dangerous to eat.

Each species of firefly has a system of signals used to send messages to each other—usually related to finding mates.

A female anglerfish has a glowing light that dangles near its mouth to attract and capture curious fish that come to investigate it.

How do deep-sea creatures make their own light?

Many deep-sea creatures produce their own light using bacteria within parts of their bodies called **photophores**. These organs go through a chemical reaction just like fireflies. Marine creatures may create light to attract mates and prey or ward off predators. Flashlight fish, for example, fold down the skin below each eye to reveal pouches of glowing bacteria, kind of like a car turning on its headlights! To confuse their predators, they "blink and run," swimming one way with the lights on, then quickly covering up the light and swimming in a different direction to escape.

How do solar panels work?

Using the Sun's light energy to create electricity is a clean, renewable way to power our homes. Solar panels contain smaller solar cells made from silicon, a naturally occurring chemical. A solar cell is a small square approximately 6 by 6 inches (15.2 x 15.2 cm) that is connected with other cells to create a solar panel. One solar panel can contain 32 to 96 solar cells. The materials inside the solar cells produce electric currents when they are exposed to sunlight, creating electricity.

It takes approximately 12 standard-size solar cells to generate enough electricity to charge a cell phone, but many solar panels are needed to power an entire home.

Does light have color?

Most sources give off white light, which is a combination of all the **wavelengths** of light. However, we can see these wavelengths, or colors, in a rainbow. Light travels through some materials, such as glass and water, but bounces off others. When light bounces off an object, it changes direction. On a sunny and rainy day, the light from the Sun travels through raindrops, causing it to bend and separate into individual wavelengths. This causes a rainbow to occur as the light exits the raindrop.

It's impossible to reach the end of a rainbow. That's because a rainbow is an ***optical illusion****. Its location is based on the observer. As you move, so will the rainbow.*

DID YOU KNOW?

When you think of a rainbow, you probably imagine an arch or semicircle of light. However, a rainbow is actually a complete circle! From the ground, we can see only the rainbow's top portion. But from the sky, it's possible to see the full circle. Not surprisingly, pilots are the people who most often see full rainbows!

CHAPTER 3: QUESTIONS ABOUT KINETIC ENERGY

What makes a roller coaster move?

Roller coasters don't have engines, so they need a different kind of energy to help speed them around the track. It's all thanks to **potential energy**, kinetic energy, and gravity.

Have you ever noticed that the first hill on a roller coaster is usually the biggest? As the train is pulled to the top of the hill, it has a lot of potential energy. When it goes over the top of the hill and is released, the potential energy becomes kinetic energy as gravity pulls it into motion. As the train climbs and falls over the length of the roller coaster, potential energy changes into kinetic energy and back again throughout the ride. There is maximum kinetic energy at the bottom of a hill. Each time the train begins to climb again, the kinetic energy changes back into potential energy.

By the time the train reaches the end, it is almost empty of kinetic energy, causing it to slow down. Specially designed braking systems built into the track bring the train to a stop—and can also stop the ride immediately if there is an emergency.

Roller coaster engineers create careful designs to maximize the use of different types of energy to power the ride and give riders a thrill.

DID YOU KNOW?

You can travel upside down! Have you wondered how a roller coaster can go upside down without its riders falling out? Inertia is the force that keeps you in place when the coaster is upside down. Although gravity pulls us toward the ground, the force of acceleration is stronger than gravity at the top of the roller coaster loop. That means you are pushed upward into your seat by acceleration rather than pulled out by gravity.

How does energy move from one object to another?

All energy is either stored or used. In order for energy to be used, it must change from potential energy to kinetic energy. Bowling is a good example of this transfer of energy. In your hands, the bowling ball isn't moving, but it has potential energy.

The force of throwing the ball causes it to move, which creates kinetic energy. When the heavy bowling ball strikes any of the pins, energy is passed to them, causing them to fall down. The faster the ball moves, the more kinetic energy it has.

Even if the bowling ball only hits one pin, the rest could fall down thanks to the transfer of kinetic energy!

Can you really start a fire with two sticks?

Starting a fire with just two sticks might sound like something only done in ancient times—but it is possible thanks to the power of kinetic energy. When the moving sticks rub against each other, **friction** causes some of the kinetic energy to change to thermal energy. Eventually, the wood can heat up enough to create smoke and **embers** that can then be held to a material that easily catches fire, such as dead grass, to create a flame.

Creating fire by rubbing sticks can be a long process. A lot of energy is needed to make the sticks burn!

How many blades does a helicopter need to fly?

In order to fly, a helicopter must have a force called **lift**. Its rotor blades, which act as wings, are the parts that provide that lift. A helicopter needs an engine and two to eight of these blades, also called airfoils. They have curved edges on top and flat sides underneath. As the blades rotate, or spin, **air pressure** is created by the kinetic energy of the moving blades. The wind passes more quickly over the curved top of the blade. That means there is less air pressure on top, pulling the helicopter upward. At the same time, the flat bottom of the blade takes on more air pressure, which creates lift from below. Without these blades, a helicopter could not fly. In fact, the blades mean a helicopter can do things an airplane cannot. It can move straight up and down, and take off or land without a runway.

Heavier helicopters require more lift, so they may have a greater number of blades.

DID YOU KNOW?

A helicopter has been flown on Mars! NASA's Ingenuity Mars Helicopter was the first controlled flight on another planet, lifting into the skies above Mars in April 2021. It successfully completed 72 flights in less than three years. The helicopter was designed to be very lightweight so that it could fly in the thin atmosphere on Mars. Its blades are much larger and spin much faster than a helicopter on Earth. Scientists tested it in extreme cold temperatures to make sure it could function on the red planet.

Can you light a match more than once?

In order to light a fire quickly, using a match is a fast option. But what if you only had one match left? Could you light it again another day? The answer lies in how a match is designed to create fire.

When a match is rubbed against a hard surface, friction causes it to heat up quickly. The head of a match is coated with a substance that, when heated, reacts with air to create a flame. As the match burns down, it turns to ash, making it impossible to light again.

Matches provide a quick way to light a fire, but they are designed to be used only once.

How does a toy jack-in-the-box work?

Have you ever been startled by the quick pop of a jack-in-the-box? This toy contains a small flexible doll that is attached to a spring fastened to the bottom of the box. When the lid of the box is closed, the spring is forced down, causing it to have a lot of **tension**. The tension in the spring causes potential energy that is stored temporarily inside. When the attached handle is turned a certain number of times, it releases the spring, opens the lid, and the jack-in-the-box pops up!

The first jack-in-the-box was made by a German clockmaker in the 1500s!

How can electricity be created from wind?

Have you ever used a fan to cool off? A fan that is plugged into an outlet uses electrical energy that flows through a cord. The electricity provides the power to spin the fan's blades, moving the air to create wind.

However, when we use wind to create electricity, we begin with the wind as the power source. The wind is the force that pushes on the blades of huge fan-like structures called turbines. The movement of the blades powers a **generator** that produces electricity. Often several turbines are grouped together to create a wind farm, which generates a lot of electricity at the same time.

Wind energy is a renewable resource, making it a great choice for generating power since it never runs out. It is also a clean energy source because it doesn't produce **carbon emissions**, which contribute to global warming, or the gradual increase of Earth's average temperature.

In the U.S., wind turbines provided around 10 percent of total electricity in 2022.

DID YOU KNOW?

Turbines are most effective at sea! Wind power is strongest over the oceans, where the wind can blow over the smooth surface of the ocean without anything to block it. So offshore wind farms have huge potential to create a lot of energy. Scientists map the oceans to measure wind speed and identify where wind farms should be built. To harness more wind power, turbines at sea are much larger than those on land.

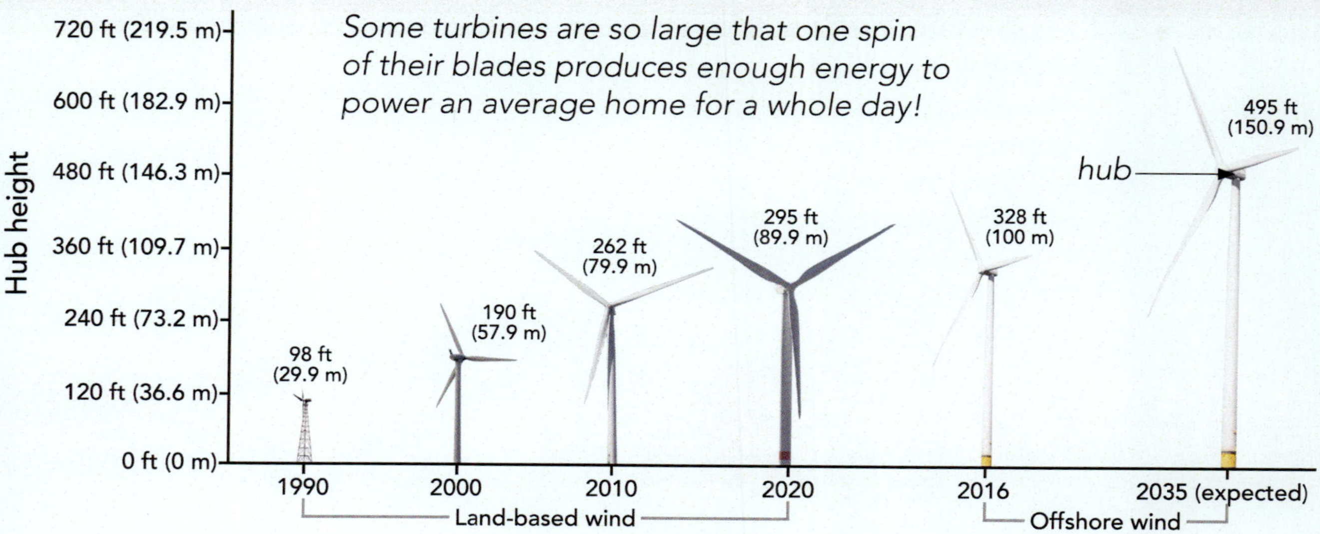

Some turbines are so large that one spin of their blades produces enough energy to power an average home for a whole day!

What causes lightning?

Lightning is caused by **electrical charges** in the atmosphere, usually during a thunderstorm. Have you heard the saying "opposites attract?" Lightning is caused when the negative electrical charge in a cloud meets the positive electrical charge on the ground below.

During a thunderstorm, ice crystals and hail form within the clouds, colliding with each other and creating electrical charges. The air carries positively charged ice crystals upward while the lower part of the cloud remains negatively charged. This then causes the ground beneath the cloud to become positively charged. The positive charge on the ground moves upward through tall objects on the ground, such as buildings and trees. Then, a flash of lightning! The opposing positive and negative charges travel and meet.

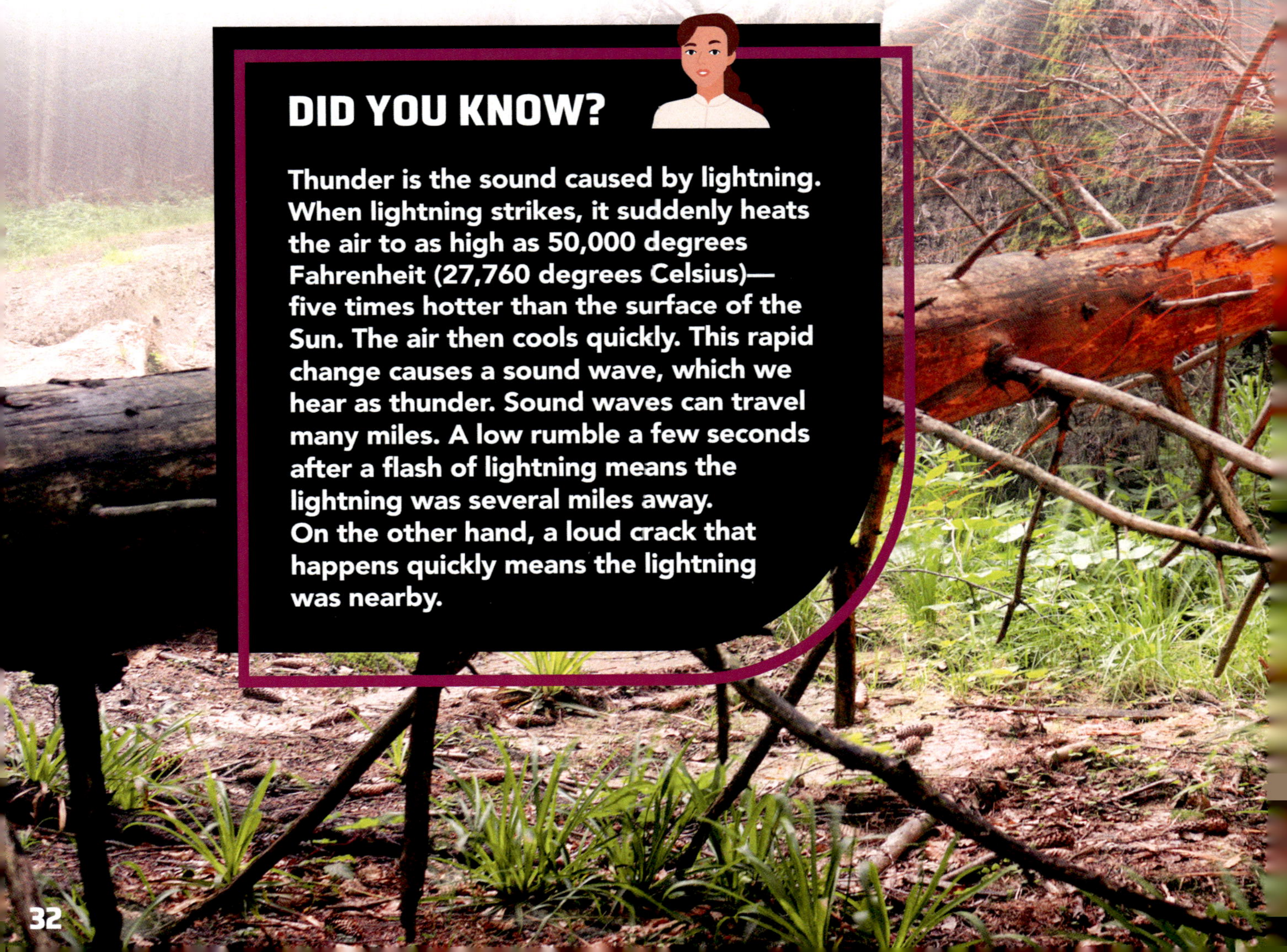

DID YOU KNOW?

Thunder is the sound caused by lightning. When lightning strikes, it suddenly heats the air to as high as 50,000 degrees Fahrenheit (27,760 degrees Celsius)—five times hotter than the surface of the Sun. The air then cools quickly. This rapid change causes a sound wave, which we hear as thunder. Sound waves can travel many miles. A low rumble a few seconds after a flash of lightning means the lightning was several miles away. On the other hand, a loud crack that happens quickly means the lightning was nearby.

Lightning often strikes trees, as they are often the tallest objects around. An estimated 20,000 fires in the U.S. are caused by lightning each year.

Why can birds stand on power lines without getting electrocuted?

Electricity travels in a **circuit**, meaning it finds another object to travel through as it returns to its source. It always follows the path of least resistance, flowing easily through objects called **conductors**. Power lines are conductors, giving electricity an easy path to flow to our homes.

Birds are not good conductors. Perched on the wire, they also do not touch anything else that could complete a circuit. So the electricity in the wire ignores the bird and flows right by. However, if a bird touched the ground, or another power line, a circuit would be created. The electricity in the wire would flow through the bird and give it a deadly shock.

Electricity in a power line will take the easiest route from the high-voltage area (the wire) to a low-voltage area (the ground). A person working on a power line wears protective clothing to make sure electricity cannot flow through them to the ground.

Does an electric eel really produce electricity?

A fish that produces electric shock waves sounds like science fiction, but it does exist! Electric eels have special organs that produce electricity. They can use an electrical shock to stun prey, such as fish or shrimp, or protect themselves from predators. After shocking its prey, an electric eel curls its body around it and gulps it down.

Most electric eels can create a shock of more than 600 volts, five times stronger than the voltage in electrical outlets in North America. In 2019, however, scientists discovered a new species of electric eel, E. voltai, *that can create a shock of up to 860 volts!*

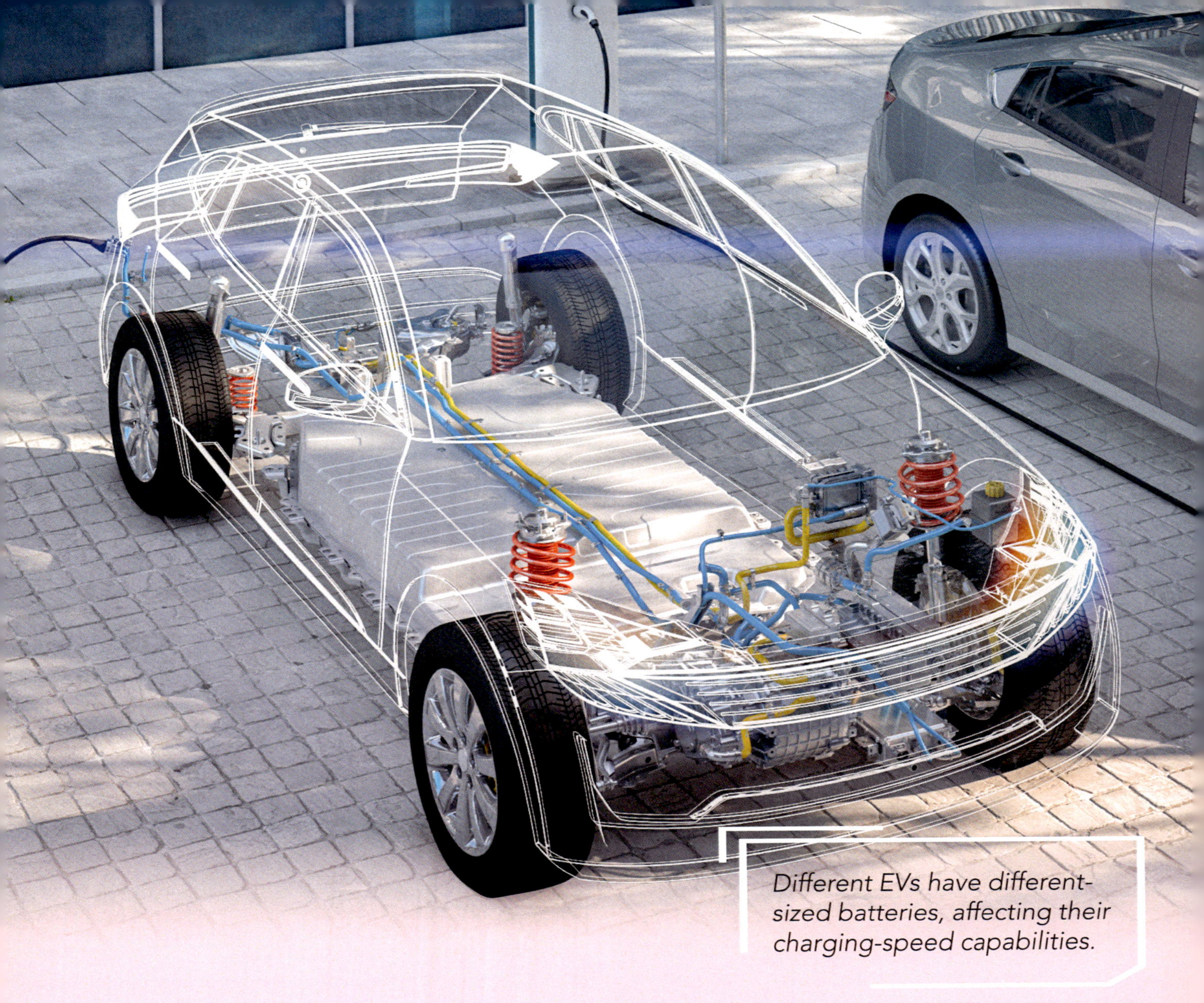

Different EVs have different-sized batteries, affecting their charging-speed capabilities.

How do electric cars work?

Electric vehicles, or EVs, don't use gasoline. Instead, they run on the energy stored in a battery. The battery is rechargeable, meaning the same battery is used over and over. An EV battery stores energy that is transferred to motors which move the wheels. Instead of filling a gas tank for power, an EV's battery must be charged regularly with a charging cord plugged into a power source.

Electric vehicle chargers work at a variety of speeds. Overnight chargers are the slowest, taking about eight to 10 hours to fully charge the battery. Fast chargers that can be found in parking garages and supermarkets take between three to four hours. Rapid chargers are the fastest, taking around 30 minutes to an hour, but in order to use rapid chargers the vehicle must have rapid-charging ability.

DID YOU KNOW?

Electric cars were invented long before gas-powered cars! Inventors such as Thomas Davenport in Vermont, U.S., started building motors powered by electricity in the 1830s. Around the same time, Scottish inventors built vehicles powered by early, non-rechargeable batteries. In 1884, an inventor in England named Thomas Parker helped develop the first commercial electric automobiles—two years before the first gasoline-powered cars. He also invented a generator to charge the battery.

Thomas Parker (middle) became interested in eco-friendly transportation because of the large amount of air pollution in London, England.

These EVs are plugged in at charging stations to recharge the batteries.

CHAPTER 5: QUESTIONS ABOUT SOUND ENERGY

How can sound break glass?

As sound energy travels from a source, it causes **molecules** to bounce around and bump into each other, creating a sound wave of fast or slow vibrations. It's hard to imagine how the vibrations in sound waves could have enough force to shatter glass, but it's possible!

Every object has a natural frequency, or the speed at which it will vibrate if hit. For example, a plucked guitar string will vibrate at a certain speed. When a sound wave matches the natural frequency of an object, it will vibrate more aggressively, making the sound louder. This is known as **resonance**. It will not necessarily cause a glass to break on its own. However, if that same sound wave is loud, or has a high volume, it will be too strong for the glass to resist. The intensity of the sound will actually shatter the glass!

The volume needed to break a glass is around the same level as a lawn mower.

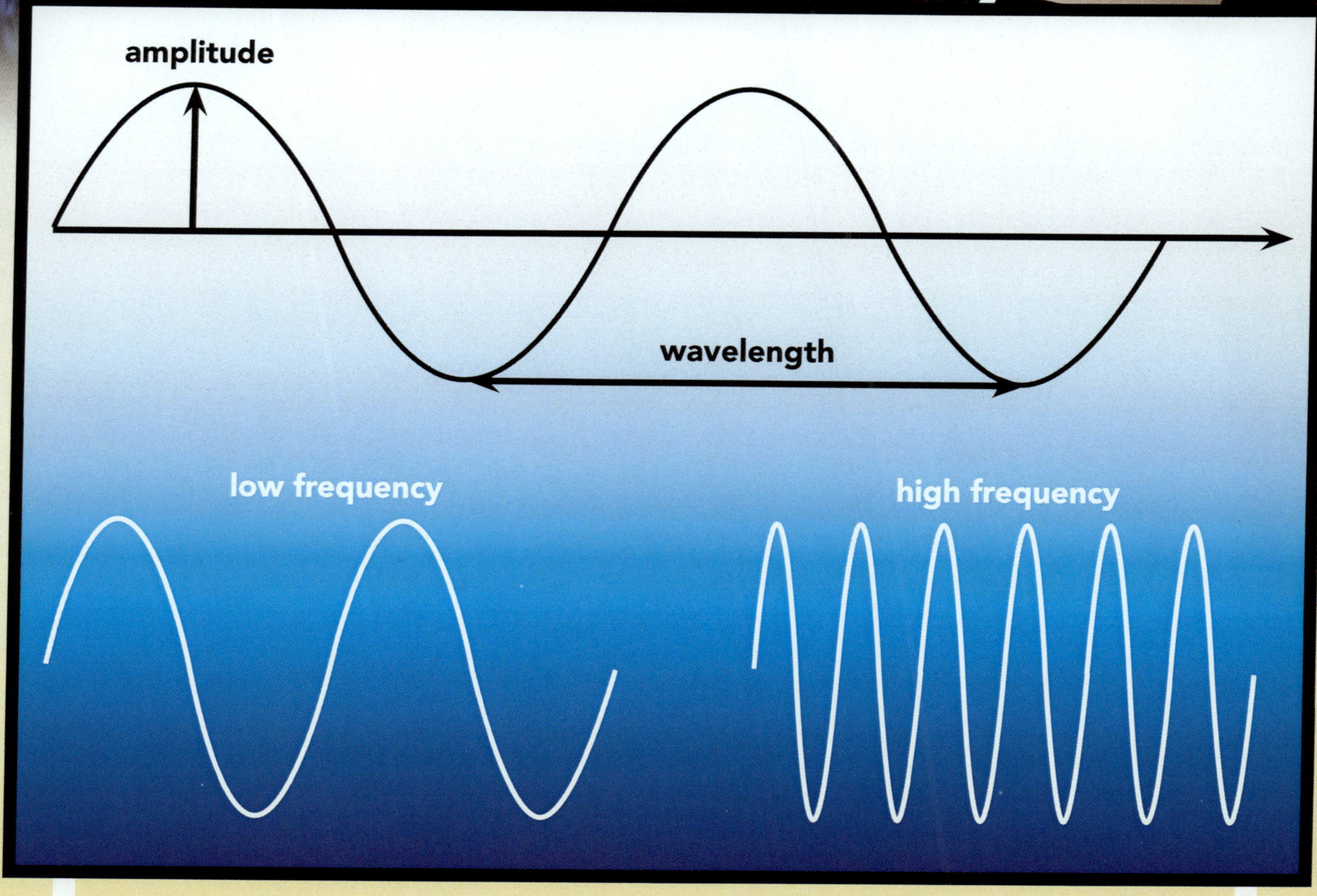

Sound waves have a frequency, amplitude, and wavelength. The frequency is the number of vibrations per second. It determines the ***pitch****. The amplitude is the height of the wave. It determines volume. The wavelength is the distance between one point on a wave to the same point on the next wave.*

Rubbing a wet fingertip around the rim of a water glass also produces sound. This causes small particles inside the glass to bounce around, transmitting vibrations into the air.

How can glasses of water create music ?

Tapping a glass of water creates vibrations, which travel in sound waves to our ears. The type of sound we hear depends on the amount of water in the glass. When there is more water in the glass, the pitch of the sound is lower. That's because more water means there is more matter for the sound to travel through, causing larger vibrations and longer wavelengths. In the same way, less water inside the glass produces faster vibrations, shorter wavelengths, and a higher pitch.

Can people feel sound?

People who can't hear can still experience sound through touch because of vibrations. Sound waves can travel through the air and be felt on the skin. They can also travel through solid materials, such as the floor or an object, and be felt when someone touches those materials.

People can even identify higher and lower-pitched notes through touch. In fact, Beethoven, one of the world's most famous **composers**, experienced hearing loss. It is said that he held a wooden stick in his teeth while resting it on his piano to feel the vibrations as he played.

Holding a balloon at a concert can help deaf and hearing-impaired people feel the vibrations of the music. That's because a balloon's thin material conducts vibrations well.

Can you hear the ocean in a seashell?

Have you ever held a seashell to your ear and thought you heard the sound of ocean waves? The sound you are actually hearing is the sound waves from the noise around you, bouncing off the inside of the shell. When you put the shell to your ear, the sound waves are captured inside and bounce off the curved walls of the shell. The types of ocean-like sounds you hear will depend on the size and shape of the seashell.

The sounds of the background noises around you sound louder when captured inside a seashell.

Why can dogs hear sounds that humans cannot?

Many animals can hear much better than humans. This is because most animals need excellent hearing to hunt prey or hide from predators.

Dogs are able to hear much higher-pitched sounds than humans can. They can thank their specially shaped ears for this ability. Large outer ears funnel sound waves into the inner ear, where the tiny bones are shaped to amplify sounds better than in humans. To amplify is to make a sound louder. This also means that dogs can hear much quieter sounds than humans can. Since dogs are related to wolves, their hearing is designed to detect even the tiniest animals.

A dog's large, curved ears work in a similar way to a seashell, capturing and amplifying sound waves.

DID YOU KNOW?

Dogs don't have the best ears! Believe it or not, cats can hear even higher pitches than dogs. Their ears are able to rotate up to 180 degrees to catch and amplify sounds. While dogs can hear sounds at frequencies up to three times higher than a human, cats can hear sounds up to four times higher! However, the animal that can hear the highest frequency of all is the greater wax moth. It can detect frequencies up to 15 times higher than a human can!

What makes an echo?

Have you ever yelled into an open space, only to hear your own voice float back to you? An echo is caused when sound waves travel through the air and bounce off of hard surfaces, back toward the source of the sound.

While soft surfaces such as cloth absorb sound, hard surfaces such as concrete reflect it. Imagine an empty room with four walls, or a cave or canyon. Sounds bounce back and forth between the hard surfaces, causing an echo. With each bounce, the echo gets quieter and will eventually stop as the sound waves run out of energy.

Echolocation is the process of using reflected sound to find objects. Some animals, such as dolphins and bats, make sounds that echo off of the body of the animal they are hunting, showing them where to find prey. Echolocation has even been used by humans! Daniel Kish, who has been blind since childhood, taught himself to locate large obstacles by creating clicking noises with his mouth and listening for the sound to bounce back. Now, he is an inspirational speaker and teaches this strategy to other people, especially children, who are blind.

Kish even used echolocation to learn how to ride a bike safely through his town!

After yelling into a large area, such as a canyon, there can be some delay before you hear the echo. That's because the sound waves have farther to travel to bounce back to your ears.

Glossary

air pressure (air PRESH-er): The weight of gases in the air

atmosphere (AT-muhs-feer): The layers of gases surrounding a planet

carbon emissions (KAHR-buhn ih-MISH-uhns): Releases of carbon substances, such as carbon dioxide and methane, into the atmosphere. Carbon emissions are a major contributor to global warming.

circuit (SUR-kit): A closed, complete path for an electrical current

commercial (kuh-MUR-shuhl): Able to be bought and sold

composer (kuhm-POH-zer): A person who writes music

conductors (kuhn-DUHK-ters): Objects or material into which electricity easily flows

electrical charge (ih-LEK-tri-kuhl chahrj): A property of matter that occurs when the number of protons and electrons differ in an atom, causing objects to attract or repel each other

embers (EM-bers): Glowing pieces of wood from a fire

fossil fuels (FOS-uhl FYOO-uhls): Fuels made from plant or animal remains

friction (FRIK-shuhn): A force that resists the movement of one object over another

generator (JEN-uh-rey-ter): A device that changes other forms of energy into electricity

gravity (GRAV-i-tee): The force that pulls objects toward Earth's center, or that pulls two objects with mass toward each other

inertia (ih-NUR-shuh): A property of matter in which an object's motion stays the same until another force changes it

kinetic (ki-NET-ik): Related to movement

lift (lift): The force created by differences in air pressure

matter (MAT-er): Any substance that has mass and takes up space

meltwater (MELT-wot-er): Water that comes from melted ice or snow

molecules (MOL-uh-kyools): Groups of two or more atoms that make up the smallest units of substances

optical illusion (OP-ti-kuhl ih-LOO-shuhn): Images that appear differently to the eye than in reality

orbit (AWR-bit): The path of one space object around another

oxygen (OK-si-juhn): A colorless, odorless gas that living things need to survive

particles (PAHR-ti-kuhls): Small pieces of matter

photophores (FOH-tuh-fawrs): Light-emitting organs in certain animals

photosynthesis (foh-tuh-SIN-thuh-sis): The process in which plants use sunlight, water, and carbon dioxide to create oxygen and energy in the form of food

pitch (pich): The quality of a sound that makes it higher or lower

potential energy (puh-TEN-shuhl EN-er-jee): Stored energy within an object

projecting (pruh-JEKT-ing): Causing an image to appear on a surface

renewable (ri-NOO-uh-buhl): Able to be used again and again without running out

resonance (REZ-uh-nuhns): The quality of vibrating or moving rapidly to make a full, clear sound

retina (RET-n-uh): The layer at the back of the eye that sends signals to the brain

rural (ROOR-uhl): Related to the country

solar retinopathy (SOH-ler ret-n-OP-uh-thee): Damage caused by the Sun that affects someone's eyesight

starch (stahrch): A type of carbohydrate found in many foods

tension (TEN-shuhn): A pulling force that is placed on an object

urban (UR-buhn): Related to cities

vibrates (VAHY-breyts): Moves back and forth very quickly

wavelength (WEYV-lengkth): The distance of a sound or light wave

Index

Comprehension Questions

1. **True or False:** In the United States, wind turbines provided around 15 percent of total electricity in 2022.
2. **The Sun is ________ away from Earth.**
   a) 96 million miles (154 million km)
   b) 91 million miles (146 million km)
   c) 93 million miles (150 million km)
3. **True or False:** Inertia is the force that keeps riders in place when a roller coaster turns upside down.
4. **________ is the height of a sound wave and determines the sound's volume.**
   a) Wavelength
   b) Amplitude
   c) Frequency
5. **Throwing a bowling ball creates ________ energy.**
   a) potential
   b) thermal
   c) kinetic
6. **True or False:** The surface of the Sun is hotter than lightning.
7. **________ produce light to confuse predators.**
   a) Flashlight fish
   b) Fireflies
   c) Anglerfish
8. **True or False:** Electricity follows the path of least resistance.

1. False, 2. c, 3. True, 4. b, 5. c, 6. False, 7. a, 8. True

About the Author

Writer and teacher Kylie Burns has written more than 30 books on a variety of topics, including animal habitats, STEM, sports, famous people, unplugged coding, and more. She lives in Uxbridge with her family and their very energetic dog!

Crabtree Publishing

crabtreebooks.com **800-387-7650**

 In Canada: We acknowledge the financial support of the Government of Canada through the Canada Book Fund for our publishing activities.

Hardcover 978-1-0398-8091-7
Paperback 978-1-0398-8452-6
Ebook (pdf) 978-1-0398-8211-9
Epub 978-1-0398-8331-4

Printed in the U.S.A./CP2025

Published in Canada
Crabtree Publishing
616 Welland Avenue
St. Catharines, Ontario
L2M 5V6

Published in the United States
Crabtree Publishing
347 Fifth Avenue
Suite 1402-145
New York, NY 10016

Written by: Kylie Burns
Series development: Janine Deschenes
Editorial coordinator: Kathy Middleton
Editor: Janine Deschenes
Proofreader: Melissa Boyce
Art direction and layout: Katherine Kantor and Margaret Salter
Photo research: Katherine Kantor
Production manager: Candice Campbell

Library and Archives Canada Cataloguing in Publication
Available at the Library and Archives Canada

Library of Congress Cataloging-in-Publication Data
Available at the Library of Congress

Images:
NASA
Goddard SDO: p. 4 (top)
JPL Caltech: p. 27 (background)
JSC: p. 9 (top)
Shutterstock
Farantsa: p. 31 (2020 turbine)
rblfmr: p. 5 (bottom)
Ruslan Lytvyn: p. 41 (background)
Wikimedia: p. 37 (top right)
NEWND: p. 31 (1990 turbine)
Oast House Archive: p. 31 (2010 turbine)
PopTech: p. 44 (bottom left)
Windkraftfan SH: p. 31 (2000 turbine)
All other images: Shutterstock